**The New Science of
Strong Materials**

The New Science of Strong Materials

or Why you don't fall through the floor

J. E. Gordon

SECOND EDITION

Princeton University Press
Princeton, New Jersey

Published by Princeton University Press, 41 William Street,
Princeton, New Jersey 08540

Copyright © 1968, 1976 by J. E. Gordon

First published 1968
Reprinted with minor revisions, 1971, 1973, 1974
Second edition 1976
First Princeton Paperback printing, 1984

LCC 83-43103
ISBN 0-691-02380

Reprinted by arrangement with Penguin Books Ltd.

Clothbound editions of Princeton University Press books
are printed on acid-free paper, and binding materials are
chosen for strength and durability. Paperbacks, while
satisfactory for personal collections, are not usually suitable
for library rebinding.

Printed in the United States of America by Princeton University
Press, Princeton, New Jersey

To my wife
for putting up with it

'*The habit of apprehending a technology in its completeness: this is the essence of technological humanism, and this is what we should expect education in higher technology to achieve. I believe it could be achieved by making specialist studies the core around which are grouped liberal studies which are relevant to these specialist studies. But they must be relevant; the path to culture should be through a man's specialism, not by-passing it . . .*

'*A student who can weave his technology into the fabric of society can claim to have a liberal education; a student who cannot weave his technology into the fabric of society cannot claim even to be a good technologist.*'
Lord Ashby, *Technology and the Academics*

Contents

List of Plates

17 Blenkinsop's rack locomotive (Science
     Museum, London)

Acknowledgements

We are grateful to the Royal Society for permission
to use the following figures: Chapter 2, Figure 5;
Chapter 3, Figure 1; Chapter 4, Figure 2; Chapter 5,
Figures 3, 4 and 5; Chapter 9, Figure 3. Figure 2,
Chapter 2, is reproduced by courtesy of J. T. Norton
and B. M. Loring (*Welding Journal, Research
Supplement, June* 1941).

Foreword to First Edition

Anyone venturing to write a book covering nearly the whole field of strong materials must give many hostages to his colleagues for his mistakes, his oversimplifications, his omissions and for his pure ignorance. Such a book is necessarily highly selective and, in practice, the choice must be personal – I hope that I may be forgiven for writing about what has interested me. There are people much better fitted than I to write, for instance, about alloy steels or titanium.

Although the book is published by permission of the Ministry of Technology, the views expressed, like the errors, are entirely my own.

Materials science and elasticity are generally pursued as rather mathematical subjects. However, I have cut out the whole of the mathematics except for a very little genuinely childish elementary algebra which can be followed by anybody with a negligible effort. In most cases I have substituted Roman titles for the Greek ones which are commonly used by the professionals (i.e. s for σ (stress), e for ε (strain) and so on). Simple transliteration returns these formulae to their common text-book form.

I should like to thank Dr W. D. Biggs of Christ's College, Cambridge, for reading the book in proof and for making useful suggestions and Mr Gerald Leach for encouragement and patient literary criticism.

Foreword to Second Edition

I owe a considerable debt of gratitude to many of my colleagues at Reading University for their help in revising this book and bringing it up to date; especially to Professor W. D. Biggs, Dr Richard Chaplin and Dr Giorgio Jeronimidis. I have also taken into account comments and criticisms from correspondents in various parts of the world who have been kind enough to write to me.

With regard to units, it is impossible to please everybody. I have tried to achieve some kind of compromise or symbiosis between S.I. and traditional Anglo-Saxon which I hope that readers will not find too emotionally disturbing.

Chapter 1   The new science of strong materials

or how to ask awkward questions

> *'Now, how curiously our ideas expand by watching these conditions of the attraction of cohesion! – how many new phenomena it gives us beyond those of the attraction of gravitation! See how it gives us great strength. The things we deal with in building up the structures on the earth are of strength (we use iron, stone and other things of great strength); and only think that all those structures you have about you – think of the "Great Eastern", if you please, which is of such size and power as to be almost more than man can manage – are the result of cohesion and attraction.'*
>
> Michael Faraday (1791–1867), *On the various forces of Nature.*

Why do things break? Why do materials have any strength at all? Why are some solids stronger than others? Why is steel tough and why is glass brittle? Why does wood split? What do we really *mean* by 'strength' and 'toughness' and 'brittleness'? Are materials as strong as we ought to expect them to be? How far can we improve existing types of materials? Could we make altogether different kinds of materials which would be much stronger? If so, how, and what would they be like? If we really could make better materials then how and where should we make use of them?

Towards the end of his life Faraday was beginning to ask some of these questions but he could make no progress in answering them and, indeed, it is only quite recently that we have been able to do so. Yet in merely asking questions about cohesion Faraday was very much ahead of his time and for many years afterwards strength and cohesion remained very unfashionable subjects with the scientific establishment.

This book is about how we have come to understand the strength of materials, about how the strengths of metals and wood and ceramics and glass and bone are interrelated and about how and why these materials behave in various kinds of structures such as machinery and tools and ships and aeroplanes and buildings and bridges.

Because what we can achieve technically has always been limited by the weaknesses of the materials of construction this new science is important. Instead of accepting our materials as something provided, arbitrarily, by Providence – as people used to until very recently – we can understand why they behave as they do and moreover, we can see much more clearly how they might be modified and improved. As a consequence, we are beginning to see our way to making radically better materials, unlike anything which has existed before, and these may open up quite new possibilities to the engineers.

Metals and non-metals

'Gold is for the mistress – silver for the maid –
Copper for the craftsman cunning at his trade.'
'Good!' said the Baron, sitting in his hall,
'But Iron – Cold Iron – is master of them all.'
Rudyard Kipling, 'Cold Iron'.

'I'll make ships out of wood; and this time I'll enjoy life and be happy.'
Weston Martyr, The Southseaman.

The great division in technology has always been that between metals and non-metals, and, although the masters like Brunel were well skilled in using both, the majority of engineers are committed to one tradition or the other. The division arises because the properties of metals and non-metals are clearly so very different – we shall presently see why – and thus the ways of using them have to be quite different. However, I am inclined to think that the mental processes involved may also be in some way opposed; metallurgists and metal engineers are apt to be practical down-to-earth people who stand no nonsense, but the non-metallurgists are probably more lyrical and imaginative.

Although perhaps we must not make too much of it, some such division has run right through the history of technology; in this book we shall review the progress of these two traditional branches of engineering in the light of modern materials science and try to see what the real problems were and why changes came when they did.

The cheapening and improvement of iron and steel during the eighteenth and nineteenth centuries* was the most important event of its kind in history – or perhaps just the most important event in history. In any case, iron and steel were particularly suited to Victorian artifacts and thus our contemporary technology is largely metallic. Metals, however, do not have a monopoly of strength. Some of the best combinations of lightness and strength are afforded by non-metals and the strongest substances in existence are the recently discovered 'whisker' crystals of carbon and of ceramics.

As the subject is developing, it now seems very possible that the coming new engineering materials will resemble much improved versions of wood and bone more closely than they will the metals with which most contemporary engineers are familiar. In the long run it is quite possible that this may affect the whole scale and character of our industries and, although it is unlikely that we shall go back to a William Morris world of woodcarvers and village carpenters or that metals will be superseded in the foreseeable future, it does make a study of the history of the use of strong materials as a whole – both metals and non-metals – relevant. Although the new techniques will be very sophisticated in many ways perhaps we may be able to get back to the patient humility of the craftsman in the face of his material which has got lost somewhere in our arrogant factories. This might result in more satisfying employment and perhaps in less industrial ugliness. If this is so, then the gain in human happiness will be very great.

We shall therefore use the modern scientific views on the strength of materials to illuminate the nature, the history and some of the applications of those materials of construction which seem to be the most important socially and technically. The selection of subjects is admittedly arbitrary and I have left out important materials, such as aluminium, where they do not seem

*The price of steel was reduced more than tenfold during the reign of Queen Victoria.

to illustrate any very interesting principle – *l'art d'ennuyer consiste à tout dire.*

The nature of materials science

It is clear that the strength of even the largest engineering structure depends in part upon chemical and physical events happening upon a molecular scale and so we shall not only have to let our ideas range freely up and down the scale of physical dimensions from the very big to the very small, but we shall also have to jump backwards and forwards from the ideas of chemistry to those of engineering. In the current phrase materials science is 'interdisciplinary'.

As soon as we start thinking about the mechanical properties of solids it becomes clear that, while we have some idea of 'how' materials behave, we have really very little idea of 'why'; naturally it is the 'why' questions which are generally the more sophisticated and the harder to answer. However, before we can tackle the reason for the way things behave we must be able to describe that behaviour accurately and objectively; this is the business of engineers. The man in the street may have rather vague views about how the solids around him deflect and break but engineers have to be precise about it and they have spent a good many generations in refining their descriptions and making them more objective. It is quite true that engineers had no idea at all why a piece of steel behaved in the way it did while a piece of concrete behaved quite differently but, at any rate, they described and measured these behaviours and wrote it all down in unreadable books. Armed with a knowledge of 'the properties' of their materials they were usually able to predict the behaviour of complicated structures – though, of course, they were sometimes wrong, in which case bridges fell down, ships sank or aeroplanes crashed.

This descriptive wisdom is embodied in the science of elasticity which defines the conditions under which structural materials receive, transmit and resist their loads; it will be necessary to have some understanding of it in order to see what the problems of strength are about. Without all the mathematics, the important principles of elasticity are really very simple but curiously difficult to understand. I think this is because most of us have grown to man's estate making use of some kind of instinctive knowledge about the strength of solids – if we hadn't we should have broken things and hurt ourselves even oftener than we did – and we think that we understand the subject intuitively and do not need telling. In fact, the real difficulty lies, not in learning about elementary elasticity, but in first getting rid of our preconceptions.

Anyone troubled with doubts on such matters is recommended to try to describe, objectively, the mechanical differences between chalk and cheese. On the whole, engineers can do this and, further, if we should wish to build a structure from either of these materials, engineers are able to predict the manner of its collapse. For the *reasons* for the differences between chalk and cheese, however, we must call on some of the other traditional divisions of science.

Solids are held together by the chemical and physical bonds between their atoms and molecules and any solid can be destroyed in several different ways, for instance, by mechanical fracture, by melting or by chemical attack. Since similar bonds have to be loosened in every case one might suppose that all these forms of dissolution were interrelated in some fairly simple way and that, now that chemists and physicists know so much about the nature of the bonds between atoms, they would have no special difficulty in explaining strength and other mechanical properties; in fact that fracture would have become practically a branch of chemistry.

As we shall see, strength is related, as of course it must be, to chemical bonding but the connexion is a roundabout one and cannot be deduced simply and entirely from classical chemistry and physics. It has turned out that not only do we need to interpret classical chemistry and physics by means of classical elasticity, but we also need to make use of rather new but important concepts such as dislocations and stress concentrations. In their day, these concepts were resisted by many of the orthodox.

It is undeniable that, until lately, the strength of materials as a science has lagged behind apparently more difficult, but perhaps more glamorous subjects and for a long time far more was known about things like wireless or the internal constitution of the stars than about what went on inside a piece of steel. In my opinion this is not so much because of the absolute difficulty of the subject but rather because of the difficulty of getting enough people in different disciplines to communicate and to take the subject seriously.

Chemists rather like to explain all the properties of matter in chemical terms and, when they had unscrambled the difficulties caused by the fact that chemists and engineers use different units to measure such things as energy, they found that their strength predictions were not only frequently a thousandfold in error but bore no consistent relationship with experiment at all. After this they were inclined to give the whole thing up and to claim that the subject was of no interest or importance anyway. Physicists did not take quite this attitude but, for a long time, most of them had other fish to fry – such as what goes on inside an atom.

Nowadays, of course, by an alliance between the physicists and the metallurgists, what occurs within a piece of metal is revealed in almost embarrassing detail, but for a long time classical metallurgy remained a descriptive science. Metallurgists knew that if one added this or that element to an alloy its properties would be affected. In the same way they knew that heating or cooling or hammering metals changed their mechanical behaviour. They could cut the metal open and observe differences in the gross crystal structure under the optical microscope but, although these differences were correlated with the behaviour of the metal, they could not be said in any way to 'explain' its behaviour at that level of explanation which we have become accustomed to demand.

Superstition and craftsmanship

Since the subject has proved so troublesome to scientists it was not to be expected that our ancestors would approach it in a very logical way and, in fact, no technical subject has been so deeply infested with superstition. A long and mostly gruesome book could, and perhaps should, be written about the superstitions associated with the making and fabrication of materials. In ancient Babylon the making of glass required the use of human embryos; Japanese swords were said to have been quenched by plunging them, red-hot, into the bodies of living prisoners. Cases of burying victims in the foundations of buildings and bridges were common – in Roman times a doll was substituted. All this was more or less in line with a good deal of primitive anthropology and seems to centre on the idea that the new structure should have a life of its own.

Latterly we have become less cruel but perhaps not much less superstitious. At any rate *some* element of irrationality about materials lingers in us all. For instance, the questions of old versus new, natural versus synthetic materials are ones which many people approach with an emotional fervour which is seldom based on real knowledge or experimental evidence. These prejudices are strongest in the non-structural applications where there is 'nothing like wool' or 'nothing like leather' but they also spread over into the structural field.

All these attitudes really amount to the idea of a kind of vitalism in materials, a 'vis-viva' on which the reliability of the substance depends; a workman will tell you that such and such has broken 'because the nature has gone out of it'. During the last war I was concerned with the supply of bamboos for making kites for anti-aircraft barrages. An importer of bamboos told me that he found it difficult to stock the lengths which we needed because they took up so much space since they had to be stored horizontally. I asked him why he did not store them vertically. 'If I did that,' he said, 'the nature would run out of the ends.'

In the past, of course, instinct and experience were the only guides to the choice of materials and to the design of structures and devices. The best traditional craftsmen were sometimes fairly good but it is a mistake to exaggerate the virtues of traditional design; the workmanship may have been excellent but the engineering design is often mediocre and sometimes shockingly bad. The wheels really did keep coming off coaches because coachbuilders were not clever enough to attach them properly. In the same way wooden ships have always leaked, quite unnecessarily, in a sea-way because shipwrights didn't understand the nature of a shearing stress and I am afraid that many of them still don't.

This excursion into the pre-scientific side of the subject might seem out of place in a book devoted to the modern science of materials, but the science of materials, like the science of medicine, has had to make its way in the teeth of a great many traditional practices and old wives' tales. Not to take account of the pit of anti-knowledge from which materials science has had to extricate itself would be unrealistic.

Atoms, chemistry and units of measurement

Even though the connexions between the strength of materials and classical physics and chemistry are not always simple or direct the subject does, of course, ultimately rest firmly on foundations of basic chemistry and physics and, for those who may have forgotten some of their 'O' level science, there is an appendix at the end of the book which tries to recap very briefly the basic minimum of physics and chemistry upon which the arguments are based. However, in the understanding of materials science it may be that an apprehension of the dimensions and scale of the various phenomena is as important as knowing the rules of chemistry and physics. In other words the 'laws' of science provide the rules of the game but the dimensions of the chess-boards – the scales upon which the games of nature and technology are played out – vary almost unimaginably. It is therefore worth spending a few moments over questions of scale and units of measurement.

Lord Kelvin used to say that one could not be said to know anything about a phenomenon until one could measure something about it and to do so naturally requires units of measurement. Latterly, S.I. units have been introduced in England, in the schools and elsewhere, but to talk entirely in terms of Newtons and metres is probably to bring in an extra difficulty or stumbling-block for most ordinary people in English-speaking countries who generally continue to think in pounds and tons and feet and inches, and so, for the larger measurements, we shall use both English and S.I. units side by side. When we come to very small measurements, however, we can all become metric and perhaps more rational. Since materials science is dealing to a considerable extent with the very small, these very small units, which are not in everyday household use, are important.

I MICRON (1μm) is $\frac{1}{10,000}$ of a centimetre, that is $\frac{1}{1,000}$ of a millimetre. The smallest thing that one can see in the ordinary way with the naked eye is roughly $\frac{1}{10}$ of a millimetre, that is about 100 microns across. The smallest thing one can see with an ordinary optical microscope is usually a little less than half a micron across. Actually the smallest thing one can see is a good deal affected by the lighting conditions. In a beam of strong light in a dark room one can see dust particles with the naked eye about 10 microns across or even less. Because one micron is around the limit of resolution in optical microscopes it is a favourite unit with biologists and other users of the optical microscope.

I ÅNGSTRÖM UNIT (1Å) is $\frac{1}{10,000}$ of a micron, that is $\frac{1}{100,000,000}$ of a centimetre.   (10Å)= 1 nanometre (nm) = 10^{-9} metre)

These are favourite units with electron microscopists and they are the units used for measuring atoms and molecules. The newer electron microscopes can see – as rather woolly blobs – particles about five Ångströms across, that is about a thousand times smaller than the best optical microscopes can achieve.* Here again the resolution is a good deal governed by the viewing conditions.

Atoms are what all matter is made of. Atoms themselves consist of a very small and heavy nucleus surrounded by a large or small cloud of planetary electrons which are waves, particles or negative charges of electricity and are very small indeed. The whole affair varies a good deal in weight and size according to the kind of atom but may be thought of as a hard but fuzzy ball very roughly two Ångströms in diameter. This is inconceivably small by ordinary standards and it is quite impossible that we should ever see individual atoms by ordinary visible light – though obviously we see them in the mass when we look at any solid.

It may be worth emphasizing that the smallest particle one can see with the naked eye is about 500,000 atoms across and the

*But then electron microscopes cost nearly a hundred times as much as optical microscopes.

smallest particle one can see with the optical microscope is about 2,000 atoms across. With the electron microscope one can see arrays of atoms in crystals, like soldiers on parade, quite easily and with a device called the field emission microscope one can see individual atoms – at least one can see that there is something which looks like a sheep in a fog on a dark evening. However if the microscope resolution were much better, as it may perhaps become, this merely raises the rather metaphysical question of what one would expect to 'see' anyway. Nothing very concrete surely?

Note There is a conversion table between English and S.I. units on page 275.

Part One

Elasticity and the theory of strength

Chapter 2   Stresses and strains

or why you don't fall through the floor

> '*It had been his custom to engage Wan in philosophical discussion at the close of each day and on this occasion he was contrasting the system of Ka-ping, who maintained that the world was suspended from a powerful fibrous rope, with that of Tai-u who contended that it was supported upon a substantial bamboo pole. With the clear insight of an original and discerning mind Ah-shoo had already detected the fundamental weakness of both theories.*'
>
> Ernest Bramah, *Kai Lung unrolls his mat.*

We are so used to not falling through the floor that we never stop to think why we don't. However the problem of how any inanimate solid is able to resist a load at all worried both Galileo (1564–1642) and Hooke (1635–1702). The understanding of simple structures and how they resist loads is a good example of a problem which, except in its molecular aspects, requires no sophisticated apparatus and could, in theory, be solved almost entirely by pure reason. This is not to say that the subject is easy; it is intellectually very difficult. The genius of Galileo and Hooke lay as much in recognizing that an important problem existed as in their contributions towards solving it, significant as these were.

As a matter of fact the general problem was probably beyond the scientific potential of the seventeenth century and it was not until well into the nineteenth that any reasonably complete idea of what was happening in a structure existed; even then this knowledge was confined to a few rather despised theoreticians. For a long time 'practical' engineers went on as they always had done, by rule of thumb. It took a long history of controversy and a series of disasters like the Tay bridge to convince these people of the usefulness of proper strength calculations.* Also it was found that reliable calculations enabled structures to be made

---

* It is said that at one time railway bridges were collapsing in the United States at the rate of twenty-five a year.

more cheaply because one could more safely economize in material. Nowadays the main difference between the qualified professional engineer on the one hand and the bench mechanic and the do-it-yourself amateur on the other lies not so much in mechanical ingenuity and skill as in an understanding of the problems of strength and energy.

Let us begin at the beginning with Newton (1642–1727) who said that action and reaction are equal and opposite. This means that every push must be matched and balanced by an equal and opposite push. It does not matter how the push arises. It may be a 'dead' load for instance: that is to say a stationary weight of some kind. If I weigh 200 pounds and stand on the floor, then the soles of my feet push downwards on the floor with a push or thrust of 200 pounds (or 900 Newtons, if you must); that is the business of feet. At the same time the floor must push upwards on my feet with a thrust of 200 pounds (or 900 Newtons); that is the business of floors. If the floor is rotten and cannot furnish a thrust of 200 pounds then I shall fall through the floor. If, however, by some miracle, the floor produced a larger thrust than my feet have called upon it to produce, say 201 pounds, then the result would be still more surprising because, of course, I should become airborne. Similarly, if a chair weighs 50 pounds, then the floor obliges by producing an upward force of exactly the 50 pounds which are needed to support the chair in its accustomed station in life. On the other hand, the force need not be a stationary weight. If I drive my car into a wall, the wall will respond by producing exactly enough force to stop the car at whatever speed it may be going, even if it kills me. Again, the wind, blowing where it listeth, pushes on my chimney pots but the chimney pots, bless them, push back at the wind just as hard, and that is why they don't fall off.

All this is merely a restatement of Newton's third law of motion which says, roughly speaking, that if the *status quo* is to be maintained then all the forces on an object must cancel each other out. This law does not say anything about how these various forces are generated. As far as the applied loads are concerned, the manner of their generation is usually straightforward: the weight of a 'dead' load arises from the action of the earth's gravitation upon the mass of the load and in the case of stopping a moving load (whether a solid, a liquid or a gas) the forces generated are those needed to decelerate the moving mass (Newton's second law of motion). The business of all structures is the conservative one of maintaining the *status quo* and in order to do this they must somehow generate adequate forces to oppose the loads which they have to carry. We can see how a weight presses down on the floor but how does the floor press up on the weight?

The answer to this question is far from obvious and the problem was the more difficult for Galileo and Hooke, in the early days of scientific thought, because the biological analogy is confusing and the tendency is, or was, to begin thinking about a problem in an anthropomorphic way. An animal has really two mechanisms for resisting loads. Its inert parts – bones, teeth and hair – resist by just the same means as any other inert solid but the living animal as a whole behaves in a quite different manner. People and other animals resist mechanical forces by pushing back in an active way: they tense their muscles and push or pull as the situation may require. If I stretch out my hand and you put a weight on it such as a pint of beer, then I have to increase the tensions in certain muscles so as to sustain the load. I am enabled to do this because the tensions in our muscles can be continually adjusted by an elaborate biological mechanism. However, the maintenance of biological tensions requires the continual expenditure of actual work (like driving a car fitted with a fluid flywheel while it is hard up against a wall – the engine is working away and using petrol and the car is pushing against the wall but neither the car nor the wall are moving). For this reason my arm muscles will sooner or later get tired and so I shall have to drink the beer to relieve them. One remains standing, not like a tripod standing inertly on the ground, but by a series of deliberate, though perhaps unconscious, adjustments of the body muscles. One gets tired standing up, and, if the muscular processes are interrupted by fainting or death, there is a dramatic collapse.

In an inanimate solid these living processes are not available. Structural materials are passive and cannot push back deliberately, so that they do not, in the ordinary sense, get tired.

They can only resist outside forces *when they are deflected*; that is, they must give way to the load to a greater or less extent in order to generate any resistance at all. By 'deflection', in this context, we do not mean that the solid moves bodily, as a whole and without changing its shape, but rather that the geometrical form of the solid is to some extent distorted so that some parts of it at least become shorter or longer by stretching or contracting within themselves. There is, and there can be, no such thing as a truly rigid material. Everything 'gives' to some extent and, as we have said, the realization that this is what structural engineering is about is what divides the professional from the amateur engineer.

When I climb a tree the deflections of the boughs under my weight will probably be very large, perhaps a matter of several inches, and are easily seen. However, when I walk across a bridge the deflections may be imperceptibly small. These are only questions of degree: there is always some deflection. Unless the deflections under loads are excessively large for the purpose of the structure they are not a fault but an inborn and unavoidable characteristic of structures with which it is the business of this chapter to come to terms. Most of us have sat in an aeroplane and watched the wing-tips going up and down. This is quite all right; the designer meant them to be like that.

It is probably obvious by this time that these deflections, be they large or small, generate the forces of resistance which make a solid hard and stiff and resistant to external loads. In other words, a solid deflects exactly far enough to build up forces which just counter the external load applied to it. This is the automatic process at the basis of all structures.

How are these forces generated? The atoms in a solid are held together by chemical forces or bonds (see Appendix 1) which may perhaps be thought of as electrical springs since there is nothing 'solid' in any crude sense to make any other kind of spring. It is these forces which bind solids together and also make the rules of chemistry. There is no distinction between the chemical bonds between atoms whose fracture yields the energy of gunpowder or petrol, and the chemical bonds which make steel and rubber strong and elastic.

When a solid is altogether free from mechanical loads (which, strictly speaking, is very seldom) these chemical bonds or springs are in their neutral or relaxed position (Figure 1). Any attempt

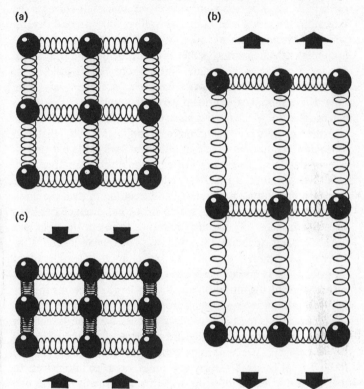

Figure 1. Simplified model of distortion of interatomic bonds under mechanical strain.

    (a) Neutral, relaxed or strain-free position.

    (b) Material strained in tension, atoms further apart, material gets longer.

    (c) Material strained in compression, atoms closer together, material gets shorter.

to push them closer together (which we call compression) or to stretch them further apart (which we call tension) involves shortening or lengthening the interatomic springs, by however little, throughout the material. This is because the middle part of the atoms may be regarded as rigid and, furthermore, in a solid the atoms do not generally exchange places – at least at moderate or 'safe' loads. The only thing, therefore, which can 'give' is the interatomic bond. These bonds or springs vary a good deal in stiffness or springiness (or, as the layman might put it, in 'strength') but most of them are much stiffer than the metal springs to which we are accustomed in ordinary life. From this it follows, of course, that the forces between the atoms are often very large indeed. We should expect this if we think about the forces which can be released by chemical explosives and fuels.

Although there is no such thing as a truly rigid solid – that is to say one which does not yield at all when a weight is put on it – in everyday life the deflections of common objects are often very small. For instance, if I take an ordinary builder's ceramic brick, stand it upright on a firm surface and tread on it, then the brick will be compressed along its length by a total distance of about $\frac{1}{50,000}$ of an inch. Any two neighbouring atoms in the brick are pushed nearer together by about $\frac{1}{500,000}$ Ångström unit (2×10^{-14}cm., or about one hundredth of a millionth of a millionth of an inch). This is an inconceivably small distance but a perfectly real movement for all that. Actually, in large structures the deflections are not by any means always tiny. In order to support their load, that is to say the roadway and the cars, the suspension cables of the Forth road bridge are permanently stretched in tension by about 0·1 per cent, or something like ten feet (three metres) in their total length of nearly two miles, or three kilometres. In this case the atoms of iron which are normally about two Ångström units apart when at rest and unloaded are kept about $\frac{2}{1,000}$ Ångström units further apart than they would be in the unstressed state.

That atoms really do move further apart when a material is stretched has been checked experimentally many times and by different methods. The most obvious way is by X-ray diffraction of stretched and unstretched specimens. The standard way of measuring the distance between the atoms in a crystal is to study the way in which an X-ray beam is deflected when it passes through the crystal. This method has been used now for sixty